NUMBER
THEORY

Authored by Mrs. Paula Burrows

Published by BSM Consulting.

ISBN: **978- 1507632574**

For orders and inquiries, please send to:

BSM Consulting
P.O. Box EE15057
Nassau, Bahamas
Email: bsmlifeconsult@gmail.com

TYPES OF NUMBERS

1. NATURAL NUMBERS
Natural numbers include all of the *counting* numbers, starting at one and continuing to infinity.

Examples: 1, 2, 3, 4, 5, 6..........

2. WHOLE NUMBERS
Whole numbers include all of the counting numbers as well as zero.

Examples: 0, 1, 2, 3, 4, 5, 6..........

3. EVEN NUMBERS
A number is even if it is divisible by *two*. If a number is not even it is odd. Even numbers have one of the following digits it the ones place: 0, 2, 4, 6, 8

Examples: -4, 0, 34, 2000, 144,298

4. ODD NUMBERS
A number is odd if it is **NOT** divisible by two. If a number is not odd it is *even*. Odd numbers have one of the following digits it the ones place: 1, 3, 5, 7, 9

Examples: 9, 21, 715, 6043

5. INTEGERS

Integers include *positive* and *negative* counting numbers.

Examples: -7, -2, 14, 45

6. RATIONAL NUMBERS

Rational numbers are numbers that can be expressed as a *fraction*.

Examples: 5, 1.75, ½

7. IRRATIONAL NUMBERS

Irrational numbers are numbers that *CANNOT* be expressed as a fraction.

Examples: $\sqrt{2}$

8. PRIME NUMBERS

A prime number is a positive whole number with exactly *two factors*, which are **one** and **itself**.

Examples: 2,7,23, 97

PRIME NUMBERS BETWEEN 1 and 100				
2	3	5	7	11
13	17	19	23	29
31	37	41	43	47
53	59	61	67	71
73	79	83	89	97

9. COMPOSITE NUMBERS

A composite number is a positive whole number with *more than* two factors.

Examples: 4, 6, 24, 36, 49.

SQUARE NUMBERS (the product of *two* similar factors)			SQUARE ROOT $\sqrt{}$ (one of two similar factors of a number)		
		Square Numbers		**Square Root**	Check
1^2	1 x 1	1	$\sqrt{1}$	1	1 x 1 = 1
2^2	2 x 2	4	$\sqrt{4}$	2	2 x 2 = 4
3^2	3 x 3	9	$\sqrt{9}$	3	3 x 3 = 9
4^2	4 x 4	16	$\sqrt{16}$	4	4 x 4 = 16
5^2	5 x 5	25	$\sqrt{25}$	5	5 x 5 = 25
6^2	6 x 6	36	$\sqrt{36}$	6	6 x 6 = 36

CUBE NUMBERS (the product of *three* similar factors)			CUBE ROOT $\sqrt[3]{}$ (one of three similar factors of a number)		
		Cube Numbers		**Cube Root**	Check
1^3	1 x 1 x 1	1	$\sqrt[3]{1}$	1	1 x 1 x1 = 1
2^3	2 x 2 x 2	8	$\sqrt[3]{8}$	2	2 x 2 x 2 = 8
3^3	3 x 3 x 3	27	$\sqrt[3]{27}$	3	3 x 3 x 3 = 27
4^3	4 x 4 x 4	64	$\sqrt[3]{64}$	4	4 x 4 x 4 = 64
5^3	5 x 5 x 5	125	$\sqrt[3]{125}$	5	5 x 5 x 5 = 125
6^3	6 x 6 x 6	216	$\sqrt[3]{216}$	6	6 x 6 x 6 = 216

TABLE OF CONTENTS

PART 1 – CORE QUESTIONS

IDENTIFY WHOLE NUMBER PLACE VALUES

EXERCISE 1

1. Complete the place value table below.

			Thousands		Tens	

2. Write the place value of the digit '**2**' in each number.

(a) 1,728

_____

(b) 23,059

_____

(c) 2,463,999

_____

(d) 23,976,104

_____

(e) 932

_____

3. Write in words:

(a) 613

_____

_____

(b) 9,327

_____

_____

(c) 10,068

_____

_____

(d) 4,397,439

_____

_____

(e) 5,009

_____

_____

4. Write down the value of:

(a) eight hundred, three tens and six ones

_____

(b) five thousand, three hundred seventy six

_____

(c) seven million five hundred ninety two thousand

_____

(d) two hundred sixteen

_____

(e) ninety nine

_____

5. Write down the number that has/is:

(a) **4** in the hundreds place, **7** in the tens place, **0** in the thousands place, **8** in the units place and **2** in the ten thousands place.

Answer

_____

(b) 100 less than the number 56,983

Answer

_____

(c) 1000 more than the number 699,237

Answer

_____

ROUND WHOLE NUMBERS

STEPS TO ROUNDING

Example: Round 53,914 to the nearest tens.

1. Locate the place value that you are rounding to.   53,9̲14

2. Consider the value of the digit immediately to the right of the digit.   53,91̲4 ←

3. If the digit in step 2 is:
   a. 5 or more, round up (i.e. add 1 to the figure in step 1)   ---
   b. Less than 5, round down (i.e. add 0 to the figure in step 1)   53,91̲4

4. Add zero placeholder(s) following rounded digit, if applicable.   53,910

EXERCISE 1

SET #1

Round to the nearest **tens.**

(a) 437 _____

(b) 512 _____

(c) 1,234_____

(d) 76,267_____

(e) 13,455_____

SET #2

Round to the nearest **hundreds.**

(a) 4,123 _____

(b) 1,012 _____

(c) 27,778 _____

(d) 101,099 _____

(e) 56,319 _____

SET #3

Round to the nearest **thousands**

(a) 5,238 _____

(b) 19,112 _____

(c) 401,736 _____

(d) 74,237 _____

(e) 3,255 _____

SET #4

Round to the nearest **ten thousands**

(a) 47,733 _____

(b) 81,042 _____

(c) 329,076 _____

(d) 509,891 _____

(e) 76,999 _____

EXERCISE 2

1. Round to the place value in brackets.

(a) 31,456 (**hundreds**) _____

(b) 421,962 (**ten thousands**) _____

(c) 7116 (**tens**) _____

(d) 98,035 (**hundred thousands**) _____

 (e) 8,089,416 (**millions**) _____

EXERCISE 3

1. Round to the place value in brackets.

(a) 523,701 (**ten thousands**)_____

(b) 1,099,982 (**hundred thousands**)_____

11

(c) 7116 (**ten millions**) _____

(d) 98,035 (**hundred thousands**) _____

(e) 8,089,416 (**millions**) _____

IDENTIFY TYPES OF WHOLE NUMBERS

EXERCISE 1

1. Identify the type of number by placing a tick under the appropriate column.

NUMBER	Number Type								
	Whole	Natural	Even	Odd	Prime	Composite	Rational	Irrational	Integers
-260									
-123									
-6									
-3									
0									
1									
2									
9									
21									
27									
29									
31									
48									
51									
91									
$\sqrt{2}$									

LIST THE SET OF FACTORS AND MULTIPLES OF A NUMBER

EXERCISE 1

1. List the **factors** of each number in the table in *ascending* order.

NUMBERS	FACTORS								
6									
10									
18									
20									
36									
50									
100									

2. List the first five non-zero **multiples** of each number in the table.

NUMBERS	MULTIPLES				
5					
7					
8					
10					
20					
25					
100					

3. Complete each statement.

(a) The smallest factor of any number is _____.

(b) The largest factor of any number is _____.

(c) The smallest multiple of any number is _____.

(d) The largest multiple of any number is _____.

3.

24	42	51
6	89	27
49	31	65

From the numbers above, write down:

(a) a multiple of 4

Answer_____

(b) a factor of 12

Answer _____

(c) a cube number

Answer _____

(d) a square number

Answer _____

(e) a prime number

Answer _____

COMPLETE NUMBER PATTERNS AND SEQUENCES

EXERCISE 1

Complete each number sequence by filling in the next three numbers.

(a) 2, 4, 8, 16, _____, _____, _____.

(b) 1, 4, 7, 10, _____, _____, _____.

(c) 7, 14, 21, 28, _____, _____, _____.

(d) 2, 20, 200, 2000, _____, _____, _____.

(e) 6, 12, 18, 24, _____, _____, _____.

EXERCISE 2

Complete each number sequence by filling in the next three numbers.

(a) 58, 68, 63, 73, 68 _____, _____, _____.

(b) 7, 8, 15, 23, 38 _____, _____, _____.

(c) 55, 34, 21, 13, 8, _____, _____, _____.

(d) 576, 288, 144, 72, _____, _____, _____.

(e) 47, 40, 45, 38, 43, 36, 41 _____, _____, _____.

EXERCISE 3

Write the next two numbers in each sequence below.

(a) 2, 3, 5, 7, 11, _____, _____.

(b) 21, 22, 24, 25, 26, _____, _____

(c) 1, 4, 9, 16, 25 _____, _____

(d) 47, 49, 51, 53, 55, 57, _____, _____

(e) 1, 8, 27, 64, _____, _____

(f) 53, 59, 61, 67, _____, _____

ADD AND SUBTRACT INTEGERS WITH AND WITHOUT THE NUMBER LINE

EXERCISE 1

1. Add the values $^+10$ to $^-10$ to the number line below.

2. Use the number line above to solve the following:

Hint: Move to the RIGHT when adding.

Move to the LEFT when subtracting.

(a) $^-4 + 8$ = _____

(b) $3 - 7 - 2$ = _____

(c) $3 - 9$ = _____

(d) $^-6 + 4 - 5$ = _____

(e) $^-2 - 6$ = _____

(f) $5 - 10 + 3$ = _____

(g) $7 - 11$ = _____

(h) $^-7 + 4 + 3$ = _____

3. Solve

(a) $3 + {}^-2$ =   _____

(b) ${}^-8 + {}^-2 - {}^-5 =$ _____

When signs and operations meet.

ADDITION SIGN MEETS

- **Positive Integer**
  e.g. $9 + {}^+2$
  $\quad 9 + 2 = 11$
- **Negative Integer**
  e.g. $9 + {}^-2$
  $\quad 9 - 2 = 7$

SUBTRACTION SIGN MEETS

- **Positive Integer**
  e.g. $9 - {}^+2$
  $\quad 9 - 2 = 7$
- **Negative Integer**
  e.g. $9 - {}^-2$
  $\quad 9 + 2 = 11$

(c) ${}^-4 - {}^-5$ =   _____

(d) $7 - 3 - {}^-5 =$ _____

(e) ${}^-8 - {}^+2$ =   _____

(f) ${}^-9 + 8 - {}^+3 =$ _____

(g) $1 + {}^-4$ =   _____

(h) $10 + {}^-6 - {}^+3 - {}^-2 =$ _____

EXERCISE 2

1. Solve **WITHOUT** using the number line.

(a) $33 - 52 =$ _____

(b) $^-25 + ^-10 =$ _____

(c) $^-47 + 51 =$ _____

(d) $73 - ^-12 =$ _____

(e) $^-98 - 42 =$ _____

(f) $57 - 12 + 10 =$ _____

(g) $19 - 24 =$ _____

(h) $^-18 + ^-12 - ^-30 =$ _____

(i) $^-56 + 39 =$ _____

(j) $76 - ^-18 + ^- 34 =$ _____

2. APPLICATION

(a) Jennifer deposited $143 in a savings account on Monday. She withdrew $59 on Wednesday and deposited $27 on Friday. How did the balance in Jennifer's bank account change and by how much?

Answer = _____

(b) Frank lost 6 pounds in July and lost 9 pounds in August. He gained 3 pounds in September. How did Frank's weight change and by how much?

Answer = _____

MULTIPLY AND DIVIDE INTEGERS

EXERCISE 1

Multiply/Divide

(a) 12 x 4 =

_____

(b) 48 ÷ ⁻4 =

_____

(c) $\dfrac{-100}{-10}$ =

_____

(d) 13 x⁻3 =

_____

(e) ⁻81 ÷ 9 =

_____

(f) 6 x⁻9 x 10 =

_____

What sign goes in the answer?		
Integer 1	Integer 2	Answer
+	+	+
+	-	-
-	+	-
-	-	+

Examples

1. 2 x 3 = 6
2. 12 ÷ -4 = -3
3. -15 ÷ -3 = 5
4. -7 x -9 = 63

(g) $^-8 \times {}^-5 =$

_____

(h) $(14 \times {}^-4) \div {}^-2$

_____

(i) $^-15 \times {}^-5 \times 3$

_____

(j) $(12 \div {}^-6) \times 10$

_____

EXPRESS A NUMBER AS A PRODUCT OF ITS PRIME FACTORS

EXERCISE 1

1. Express each of the following numbers as a product of its prime factors (PRIME FACTORISATION).

(a) 18

Answer _____

(b) 24

Answer _____

(c) 72

Answer _____

(d) 144

Answer _____

(e) 200

Answer _____

(f) 169

Answer _____

EXERCISE 2

Complete each FACTOR TREE by filling in the missing numbers correctly.

(a)

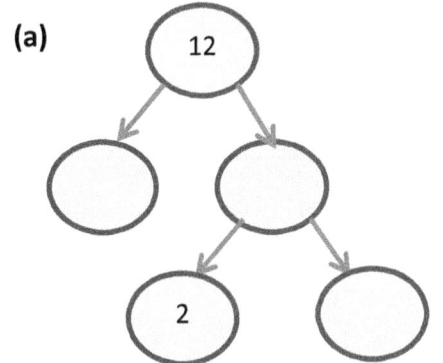

(b)

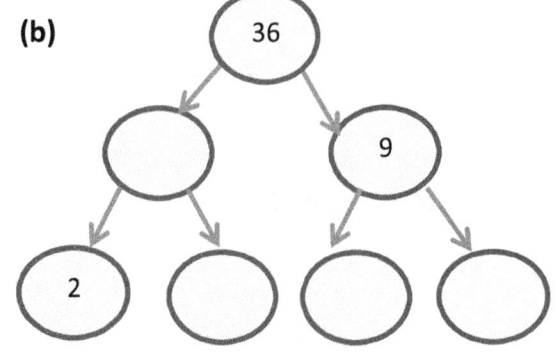

(c)

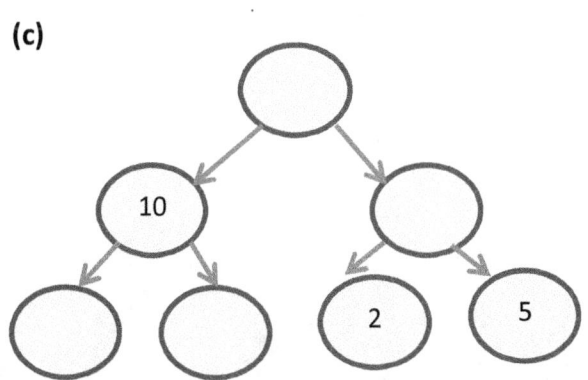

(d)

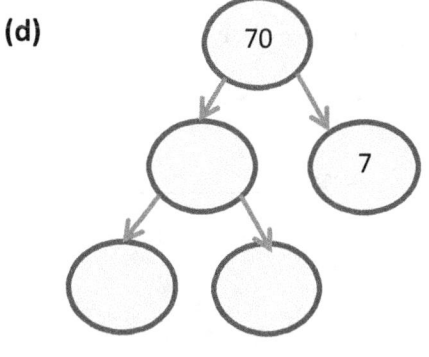

CALCULATE THE LCM AND HCF/GCF OF A SET OF NUMBERS

EXERCISE 1

1. Calculate the LCM of the following numbers:

(a) 4 and 6

Answer _____

(b) 4 and 8

Answer _____

(c) 6, 9 and 18

Answer _____

(d) 8, 12 and 24

Answer _____

(e) 5 and 6

Answer _____

(f) 5, 10 and 25

Answer _____

(g) 12, 30 and 60

Answer _____

(h) 2, 3 and 5

Answer _____

EXERCISE 2

1. Calculate the HCF of the following numbers:

(a) 6 and 12

Answer _____

(b) 24 and 36

Answer _____

(c) 10, 20 and 150

Answer _____

(d) 31 and 37

Answer _____

(e) 8, and 12

Answer _____

(f) 30 and 36

Answer _____

(g) 20, 30 and 40

Answer _____

(h) 5, 11 and 19

Answer _____

EXERCISE 3

1. The prime factors of two numbers are listed below.

2 x 2 x 3 x 3 x 5
2 x 2 x 2 x 3 x 5

(a) Write down the Highest Common Factor (HCF) of these numbers.

Answer _____

2. Use prime factorisation to calculate the **LCM** of each set of numbers.

(a) 4, and 10

Answer _____

(b) 8, 20 and 40.

Answer _____

CALCULATE SQUARE AND CUBE ROOTS USING PRIME FACTORIZATION

EXERCISE 1

1. Solve using prime factorization.

(a) $\sqrt{144}$

Answer _____

(b) $\sqrt{400}$

Answer _____

(c) $\sqrt{225}$

Answer _____

(d) $\sqrt{196}$

Answer _____

(e) $\sqrt{484}$

Answer _____

(f) $\sqrt{625}$

Answer _____

EXERCISE 2

1. Solve using prime factorization.

(a) $\sqrt[3]{8}$

Answer _____

(b) $\sqrt[3]{125}$

Answer _____

(c) $\sqrt[3]{216}$

Answer _____

(d) $\sqrt[3]{512}$

Answer _____

(e) $\sqrt[3]{1000}$

Answer _____

(f) $\sqrt[3]{729}$

Answer _____

EXERCISE 3

Complete the table below.

Number		SQUARE	SQUARE ROOT
(a)	4		
(b)	9		
(c)	16		
Number		CUBE	CUBE ROOT
(d)	8		
(e)	1		
(f)	1000		

Exercise 4 –Review

1. The prime factors of 28, 42 and 70 are given below.

$$28 = 2 \times 2 \times 7$$
$$42 = 2 \times 3 \times 7$$
$$70 = 2 \times 5 \times 7$$

Use the factors to write down the HCF/GCF of the given numbers.

Answer _____

2. Write down the value of 'x' in each case.

(a) $\sqrt{x} = 12$

Answer _____

(b) $\sqrt{x} = 10$

Answer _____

(c) $\sqrt{x} = 8$

Answer _____

(d) $\sqrt{x} = 5$

Answer _____

ADD, SUBTRACT, MULTIPLY AND DIVIDE WHOLE NUMBERS

EXERCISE 1

(a)   348
    + 567
    _____

(b) 1489
    - 1225
    _____

Answer (a)_____

Answer (b)_____

(c) 674
    x 5
    _____

(d) 4) 1244

Answer (c)_____

Answer (d)_____

36

EXERCISE 2

(a)   170
      +   5
      __34__
      _____

(b) 9875
     - 2845
     _____

Answer (a)_____   Answer (b)_____

(c)   439
      x 6
      _____

(d) 9⟌723

Answer (c)_____   Answer (d)_____

PART II - EXTENDED QUESTIONS

EVALUATE PROBLEMS CONTAINING INDICES

EXERCISE 1

Write the value of:

(a) $2^3 + 7$

(b) $3^3 + 1$

Answer (a) _____

Answer (b) _____

(c) $2^3 \times 3^2$

(d) $5^2 + 3^2$

Answer © _____

Answer (d) _____

EXERCISE 2

Evaluate:

(a) $\sqrt{100} - 2^3$

(b) $\sqrt[3]{64} + \sqrt{25}$

Answer (a) _____

Answer (b) _____

(c) $4^3 \div \sqrt{64}$

(d) $\sqrt{144} - 2^4$

Answer (c) _____

Answer (d) _____

SOLVE PROBLEMS INVOLVING MULTIPLE OPERATIONS (ORDER OF OPERATION)

EXERCISE 1

Use PEMDAS (or BODMAS) to solve.

(a) 4+3 x 2   (b) 9 - $\sqrt{144}$

Answer (a) _____   Answer (b) _____

(c) 12÷6 + 5 x 2   (d) $\sqrt[3]{64}$ + (18÷9)^2

Answer (c) _____   Answer (d) _____

EXERCISE 2

Use PEMDAS (or BODMAS) to solve.

(a) $7 \times 3 - 4^2$

(b) $\sqrt{81} + 4 \times 5$

Answer (a) _____

Answer (b) _____

(c) $(20 \div 4) - 6 \div 3$

(d) $10 \times (28 \div 7)^3$

Answer (c) _____

Answer (d) _____

EXERCISE 3

Evaluate:

(a) $3^2 + 5^2$

Answer _____

(b) (i) $(6,319 - 486) \div 8$

Answer _____

(ii) Write your answer to part (b)(i) correct to the nearest hundreds.

Answer _____

SOLVE APPLICATION PROBLEMS INVOLVING WHOLE NUMBERS/INTEGERS

EXERCISE 1

Solve:

1. Kimmy weighed 215 pounds. She started a diet. The table below shows the results over the first four weeks.

WEEKS	1	2	3	4
Weight gained	0 pounds	3 pounds	0 pounds	2 pounds
Weight Lost	4 pounds	0 pounds	5 pounds	0 pounds

(i) How much does Kimmy weigh at the end of the fourth week?

Answer _____

2. A school's playground measuring 35 metres, 47 metres, 60 metres and 32 metres, is enclosed by a wooden fence. How many metres of fencing are used?

Answer _____

3. Mr. Jones earns $45,733 per year.

How much money does he earn in one month?

Answer _____

Write your answer in (a) correct to the nearest dollar.

Answer _____

4. The sum of three numbers is 166. Two of the numbers are 64 and 39. What is the third number?

Answer _____

5. (i) List ALL of the prime numbers found on the balls below.

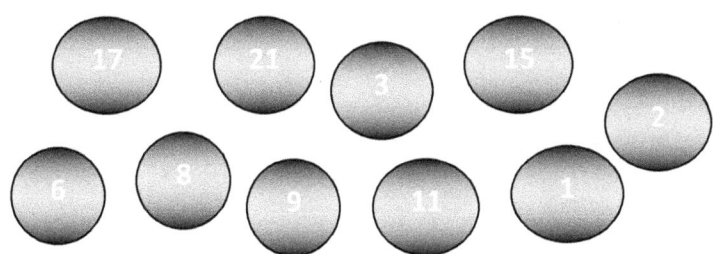

Answer _____

(ii) Calculate the **sum** of the prime numbers in (i).

Answer _____

6. Multiply 367 by 8.

.

Answer _____

7. How many times is 23 contained in 391?

Answer _____

8. 5 boxes have a total weight of 800 kg. One box weighing 120 kg is taken away.

(i) Calculate the **average weight** of the remaining 4 boxes.

Answer _____

9. Fill in the missing numbers in the square below so that the totals for each row, column and diagonal are the same.

9		16
	10	
		11

10. An ice-cream shop sold 153 cones on Monday, 268 cones on Tuesday, 410 cones on Wednesday and 308 cones on Thursday. How many cones must be sold on Friday to bring the total up to 1638 cones?

Answer _____

This book is written for Junior students who have challenges with Mathematics. The concept of 'practicing' builds students' confidence in approaching various Mathematics subjects. The result is 'success'. This book was written by Paula Burrows and inspired by her time in the classroom teaching Mathematics at the junior level.

Practicing Number Theory for Success is the first book in the *Practicing Series*. Other books in this series include:

- Practicing Geometry for Success
- Practicing Measurement for Success
- Practicing Mensuration for Success
- Practicing Algebra for Success
- Practicing Sets for Success
- Practicing Symmetry and Graphs for Success
- Practicing Fractions for Success
- Practicing Decimals for Success
- Practicing Percentages for Success
- Practicing Ratio and Proportions for Success
- Practicing Probability and Statistics for Success

The books in this series are also accompanied by their respective answer keys.

For a comprehensive version of all these topics, please see

Practicing Mathematics for BJC Success

It contains practice questions on all the topics listed in the Practicing Series.

www.ingramcontent.com/pod-product-compliance
Lightning Source LLC
Chambersburg PA
CBHW080650180526
45168CB00008B/3370